A NEW ORDER OF FISHLIKE AMPHIBIA FROM THE PENNSYLVANIAN OF KANSAS

AF893842

THEODORE H. EATON
PEGGY LOU STEWART

ZINC READ

Published by
This edition copyright © 2024 by Zinc Read
Design and setting By Zinc Read
Email- zincread@gmail.com
ISBN-978-93-6277-417-0

CONTENTS

INTRODUCTION

A slab of shale obtained in 1955 by Mr. Russell R. Camp from a Pennsylvanian lagoon-deposit in Anderson County, Kansas, has yielded in the laboratory a skeleton of the small amphibian *Hesperoherpeton garnettense* Peabody (1958). This skeleton provides new and surprising information not available from the holotype, No. 9976 K. U., which consisted only of a scapulocoracoid, neural arch, and rib fragment. The new specimen, No. 10295 K. U., is of the same size and stage of development as the holotype and it is thought that both individuals are adults.

The quarry, University of Kansas Museum of Natural History Locality KAN 1/D, is approximately six miles northwest of Garnett, Anderson County, Kansas, in Sec. 5, T. 19S, R. 19E, 200 yards southwest of the place where *Petrolacosaurus kansensis* Lane was obtained (see Peabody, 1952). The Rock Lake shale, deposited under alternately marine and freshwater lagoon conditions, is a thin member of the Stanton limestone formation, Lansing group, Missourian series, and thus is in the lower part of the Upper Pennsylvanian.

Peabody (1958) placed *Hesperoherpeton* in the order Anthracosauria, suborder Embolomeri, family Cricotidae. Study of the second and more complete specimen reveals that *Hesperoherpeton* is unlike the known Embolomeri in many important features. The limbs and braincase are more primitive than those so far described in any amphibian. The vertebrae are comparable to those of Ichthyostegalia (Jarvik, 1952), as well as to those of Embolomeri. The forelimb is transitional between the pectoral fin of Rhipidistia and the limb of early Amphibia. The pattern of the bones of the forelimb closely resembles, but is simpler than, that of the hypothetical transitional type suggested by Eaton (1951). The foot seemingly had only four short digits. The hind limb is not known.

The new skeleton of *Hesperoherpeton* lies in an oblong block of limy shale measuring approximately 100 x 60 mm. After preparation of the entire lower surface, the exposed bones and matrix were embedded in Bioplastic, in a layer thin enough for visibility but giving firm support. Then the specimen was inverted and the matrix removed from the opposite side; this has not been covered with Bioplastic. The bones lie in great disorder, except that some parts of the roof of the skull are associated, and the middle section of the vertebral column is approximately in place. The bones of the left forelimb are close together but not in a natural position. The tail, pelvis, hind limbs and right forelimb are missing. Nearly all the bones present are broken, distorted by crushing, incomplete and scattered out of place, probably by the action of currents. The complete skeleton, in life, probably measured between 150 and 200 mm. in length.

The specimen was studied at the Museum of Natural History, University of Kansas, with the help of a grant from the National Science Foundation, number NSF-G8624. The specimen was discovered in the slab by Miss Sharon K. Moriarty, and was further cleaned by the authors. Mr. Merton C. Bowman assisted with the illustrations. We are indebted to Dr. Robert W. Wilson for critical comments.

SKULL

DORSAL ASPECT

(Figs. 1, 2)

In reconstruction, the skull measures approximately 8.0 mm. dorsoventrally at the posterior end. The height diminishes anteriorly to about 1.5 mm. at the premaxillary. The length is about 15.5 mm. in the median line, or 24.0 mm. to the tip of the tabular, and the width about 16.0 mm. posteriorly. The snout is blunt, continuing about 1-2 mm. anterior to the external nares. Each of the tabulars has a slender posterior process 5.0 mm. long, which probably met the supracleithrum; the intertabular space is about 8.5 mm. wide. The orbits are approximately 5.5 mm. in diameter and extend from the maxillary to within about 3.0 mm. of the midline dorsally. The pineal opening is 1.8 mm. anterior to the occipital margin of the skull.

Reduction of bones at the back of the skull seems to have eliminated any dermal elements posterior to the squamosal, while enlargement of the orbit has removed most of the postorbital series, leaving the squamosal as the only cheekbone. There is apparently no jugal or postfrontal.

The squamosal of *Acanthostega* (Jarvik, 1952) is articulated under the tabular and reaches forward and down, much as if it were an opercular in reversed position. Internally, it must lie against the otic capsule below the tabular, partially concealing the stapes. The bone that we suppose to be the squamosal of *H. garnettense* is of similar shape, of about the same size and has internally an articular surface at one corner, bounded by a pair of ridges in the shape of a V. This articular surface probably fitted on a lateral process extending from the roof of the neurocranium, over the front of the otic capsule.

The premaxillary extends posterolaterally to a distance 5.5 mm. from the midline and attains a width at its broadest point of about 1.5 mm. The posterior edge is slightly concave and in part forms the anterior border of the naris.

Fig. 1. *Hesperoherpeton garnettense* Peabody. Skull, dorsal view. Postorbital processes of the neurocranium are shown in dotted outline. KU 10295, x 4.

The nasal is triangular and, with the lacrimal, forms the medial border of the naris. The length of the medial side of the nasal bone is approximately 5.0 mm., the transverse width is 3.8 mm., and the extent of the posterolateral border is 5.5 mm.

The maxillary meets the premaxillary lateral to the naris, borders the naris posteroventrally, and continues posteriorly beneath the orbit, of which it forms the external border. The maxillary is about 8.5 mm. long, and immediately anterior to the orbit has a maximum width of 1.3 mm.

The lacrimal fills the remaining rim of the narial opening between the nasal and maxillary, and extends to the anterior edge of the orbit. The length, from naris to orbit, is 4.2 mm.; the width ranges from 1.0 mm. anteriorly to 2.5 mm. posteriorly.

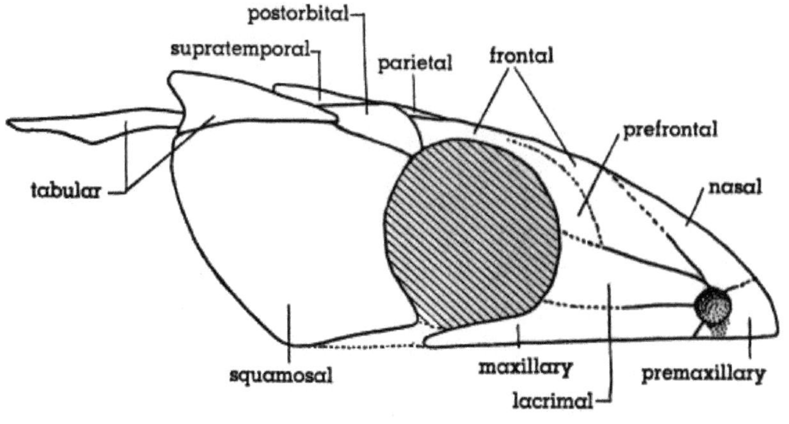

Fig. 2. *Hesperoherpeton garnettense* Peabody. Skull, lateral

view, showing relatively large orbit and absence of smaller circumorbital bones. KU 10295, x 4.

The external naris is approximately 1.0 mm. in diameter. It is slightly anterodorsal to the internal naris and 4.0 mm. lateral to the midline.

The dorsal margin of the orbit appears to be formed by the frontal. The anterior part of this margin, however, may be formed by a prefrontal, which is not clearly set off by a suture. The frontal extends 3.8 mm. in the midline, and anteriorly and laterally borders the nasal and lacrimal, respectively. A faint pattern of pitting radiates on the surface from the center of ossification of the frontal. There is also a pit indicating the presence of a supraorbital sensory pore.

The parietal bones enclose the pineal opening, approximately 2.5 mm. posterior to the suture with the frontal. The foramen is about 0.5 mm. in diameter. Laterally the parietal meets the medial angle of the postorbital and the medial border of the supratemporal. No bone of this animal shows the deep pitting and heavy ornamentation characteristic of many primitive Amphibia.

The postorbital meets the anterolateral corner of the parietal for a distance of 0.5 mm., the anterior edge bordering the frontal bone and the orbit for a combined distance of about 3.0 mm. The lateral margin is slightly convex, and is probably interrupted behind by the anterior point of the tabular. Medially, the concave margin of the postorbital meets the supratemporal for about 3.5 mm.

The supratemporal is thus wedge-shaped and located between the parietal and the postorbital. The posterior edge of the supratemporal protrudes as a convex border slightly behind the end of the parietal, and measures 3.0 mm. around the curve to the parietal suture.

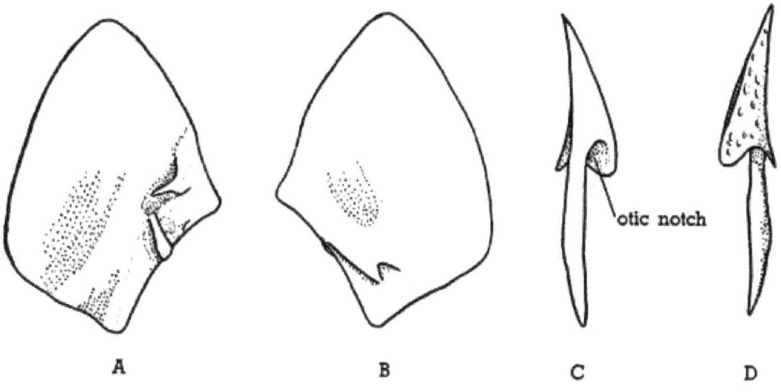

Fig. 3. *Hesperoherpeton garnettense* Peabody. A, left squamosal, internal surface. B, left squamosal, external surface. C, right tabular internal surface. D, right tabular, external surface. KU 10295, all x 4.

The squamosal (Fig. 3 A, B) is a large, somewhat rectangular bone extending from the back of the orbit to the posterior extremity of the cheek. It outlines almost entirely the posterior border of the orbit, the ventrolateral portion of the cheek region, and the lateral border of the top of the skull behind the orbit. Dorsally, the squamosal meets the anterior half of the tabular and the lateral border of the supratemporal. Near the anteroventral edge of the squamosal there is a small pit, probably related to a postorbital sensory pore in the skin.

The tabular (Fig. 3 C, D) is pointed anteriorly, where it probably fits against the lateroposterior edge of the postorbital. The dorsal part of the bone flares out and down, forming a small otic notch at a point halfway back. Posteriorly, the flange attains a dorsoventral width of 2.0 mm. at the edge of the notch. The slender posterior process of the tabular which continues beyond the flange is approximately 0.5 mm. in diameter and 5.0 mm. long.

VENTRAL ASPECT

The palatal view of the skull shows the paired premaxillary, maxillary, palatine, pterygoid, and quadrate bones. The openings for the internal nares, the ventral orbital fenestrae, and the subtemporal fossae are readily recognized. The quadrate processes extend posteriorly leaving a large gap medially at the posterior end of the skull.

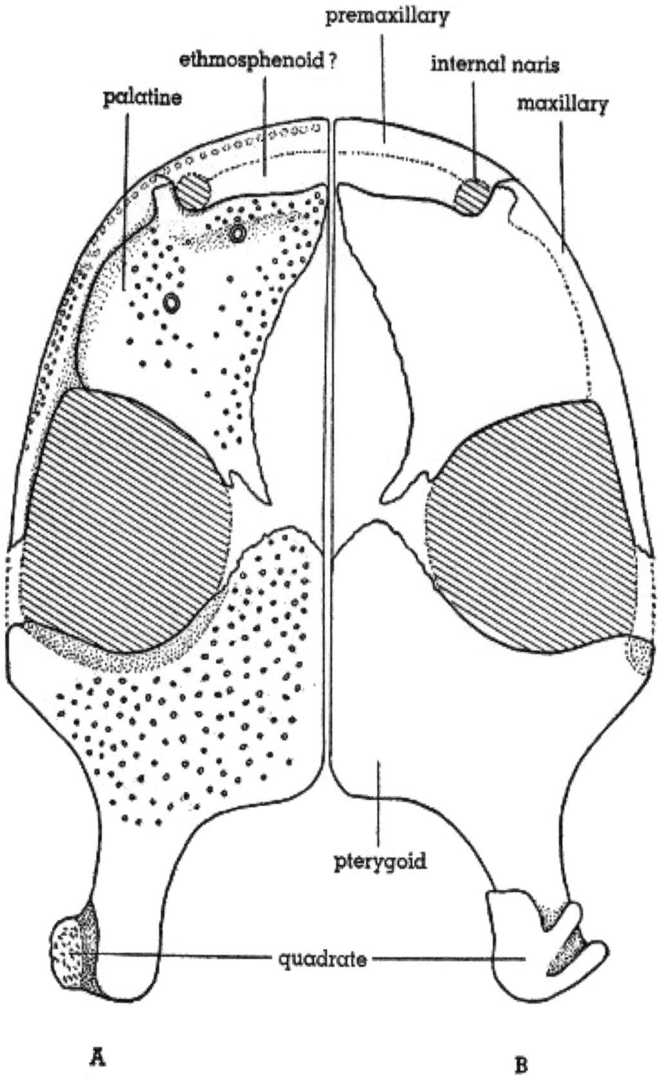

Fig. 4. ***Hesperoherpeton garnettense*** **Peabody. Palate**

reconstructed; ventral aspect at left, showing teeth, dorsal aspect at right. KU 10295, x 4.

The left quadrate appears to be in place on the posterior prong of the pterygoid. The dorsal side of the quadrate is grooved between two anterolaterally directed ridges. The groove, which probably held the end of the stapes, extends about half the width of the quadrate itself. The width of the quadrate is 4.0 mm., the length is 4.5 mm. medially and about 2.0 mm. laterally. In ventral view the quadrate appears to project laterally, but is incomplete and its shape uncertain. The distance from the posterior end of the quadrate to the visible posterior edge of the orbital fenestra, which opens ventrally, is 10.0 mm.

This region between the quadrate and the orbit is occupied by a pterygoid with three projections. Anteriorly, the pterygoid outlines most of the posterior edge of the orbit (a distance of about 6.5 mm.). A lateral process separates the orbit from the subtemporal fossa. A posteriorly directed edge defines the fossa, which extends about 6.5 mm. anteroposteriorly. The lateral process of the pterygoid terminates 10.0 mm. from the midline. Both the lateral and posterior pterygoid processes are approximately 2.0 mm. wide. The greatest width of the subtemporal fossa is about 2.0 mm. The medial border of the orbital fenestra is missing, but apparently consisted of the pterygoid for at least the posterior half.

Along the posterior edge of the orbital fenestra, there is a narrow, dorsally projecting flange of the pterygoid. The lateral opening of the orbit is approximately 7.5 mm. wide.

The remaining border of the orbital fenestra on the anterior and medial sides is formed by a bone occupying the position of palatine and vomer; for convenience we designate this as palatine. When reconstructed in its probable position in relation to the pterygoid, the left palatine lacks a section, on its medial and posterior edges, measuring about 2.5 mm. by 9.0 mm. The lateral margin of the palatine is convex; about 5.5 mm. anterior to the orbit this margin curves into a strong anteriorly pointing projection, medial to which is seen the internal narial opening. The remaining anterior edge is slightly convex, smoothly rounded, and meets the midline about 9.0 mm. anterior to the pterygoid.

The void area medial to the palatine and anterior to the pterygoid does not fit any bone which we can recognize as the parasphenoid. It is thus suspected that this area is covered in part by the missing edge of the palatine and partly by an anteromedial extension of the pterygoid. Of course a parasphenoid may also have been present.

The position, length, and shape of the premaxillary shown in palatal view (Fig. 4) are primarily based upon the dorsal appearance since ventrally most of it cannot be seen. At the point where it forms the anterior border of the internal naris, the premaxillary is slightly wider than the maxillary and seems to become narrower as it approaches the midline.

The ethmosphenoid, which we cannot identify, may have been exposed in a gap between the premaxillary and the palatine. The gap measures approximately 8.0 mm. wide and ranges up to 1.0 mm. anteroposteriorly.

The maxillary begins at a suture with the premaxillary lateral to the naris and continues posteriorly, bordering the orbit with a width of about 1.2 mm. It then tapers to a point approximately 2.0 mm. anterior to the lateral projection of the pterygoid. The width of the maxillary at this point is 0.8 mm. and the posterior end is broken; probably when complete it approached the pterygoid, and either met the latter or had a ligamentous connection with it. As nearly as can be determined, the total length of the maxillary is approximately 12.0 mm.

The teeth on the maxillary are small and seem to be in two longitudinal rows. The palatine bears two large, grooved teeth anteriorly; the first is approximately 1.0 mm. posteromedial to the naris and the second is about 3.0 mm. posterior and slightly lateral to the naris. The flat ventral surfaces of the palatine and pterygoid bear numerous small teeth distributed as shown in Fig. 4.

BRAINCASE AND OCCIPITAL REGION *(FIG. 5)*

The parts of the neurocranium are scattered, disconnected and incomplete, but it is possible to make out a number of features of the otico-occipital section with fair assurance. In posterior view the notochordal canal and foramen magnum are confluent with each other, and of great size relative to the skull as a whole. The notochordal canal measures 2.8 mm. in diameter, and the foramen magnum about 4.0 mm. The crescent-shaped supraoccipital rests on the upright ends of the exoccipitals, but between the latter and the basioccipital no sutures can be seen. Probably the whole posterior surface of the braincase slanted posteroventrally; consequently the rim of the notochordal canal was about 3.0 mm. behind the margin of the parietals.

The U-shaped border of the notochordal canal is a thick, rounded bone, comparable in appearance to the U-shaped intercentra of the vertebrae. This bone apparently rested upon a thinner, troughlike piece (Fig. 5 B) forming the floor of the braincase. The latter is broad, shallow, concave, open midventrally and narrowing anteriorly to form a pair of articular processes. Since no sutures can be seen in this structure, it probably is the ventral, ossified portion of the basioccipital. Watson (1926, Fig. 4 B) illustrates the floor of the braincase in *Eusthenopteron,* with its more lateral, anterior portion labelled prootic, but in our specimen the corresponding part could scarcely have formed the anterior wall of the otic capsule, being entirely in the plane of the floor. The two articular surfaces anteriorly near the midline suggest that a movable joint existed between the otico-occipital part of the braincase and the ethmosphenoid part, as in Rhipidistia (Romer, 1937). We have found nothing in the specimen that could be referred to the ethmosphenoid; it may have been unossified.

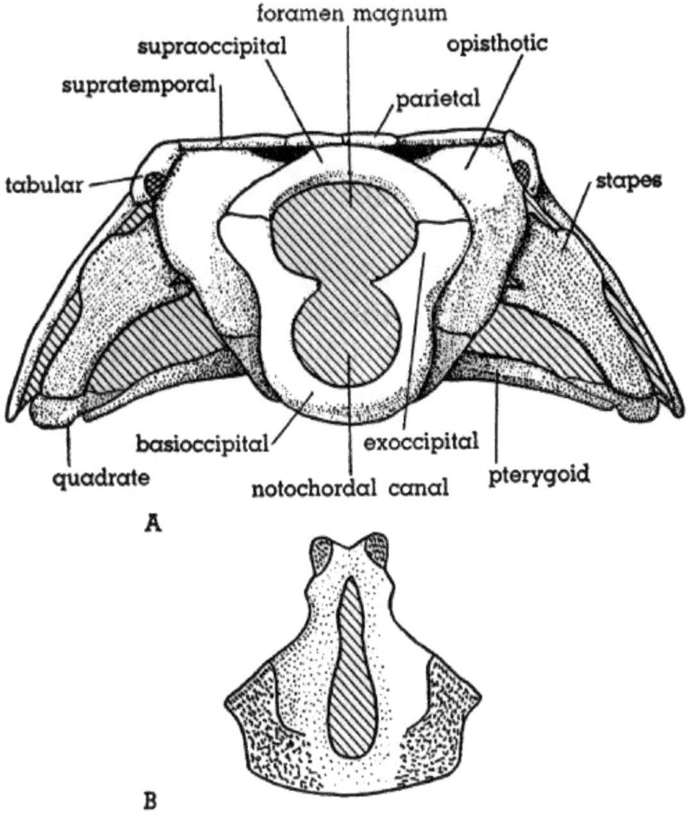

Fig. 5. ***Hesperoherpeton garnettense*** **Peabody, KU 10295, x 4. A, occipital view of skull; B, basioccipital bone in dorsal (internal) view.**

The otic capsules appear to have rested against lateral projections of the basioccipital. The single otic capsule that can be seen (the right) is massively built, apparently ossified in one piece, with a shallow dorsomedial excavation, probably the vestige of a supratemporal fossa. On the lateral face is a broad, shallow depression dorsally, and a narrower, deeper one anteroventrally; these we suppose to have received the broader and narrower heads of the stapes, respectively. The posterior wall of the otic capsule we have designated opisthotic in the figure. Anterior to the otic capsule the lateral wall of the braincase cannot be seen, and may not have been ossified.

The roof of the braincase is visible in its ventral aspect, extending from approximately the occipital margin to a broken edge in front of the parietal foramen, and laterally to paired processes which overlie the otic capsules directly behind the orbits (see dotted outlines in Fig. 1). Each of these postorbital processes, seen from beneath, appears to be the lateral extension of a shallow groove beginning near the midline. Presumably this section of the roof is an ossification of the synotic tectum. It should be noted that the roof of the braincase proper is perfectly distinct from the overlying series of dermal bones, and that the parietal foramen can be seen in both. The roof of the braincase in our specimen seems to have been detached from the underlying otic capsules and the occipital wall.

The bone that we take to be the stapes is blunt, flattened (perhaps by crushing), 5.0 mm. in length, and has two unequal heads; its width across both of these is 4.0 mm. The length is appropriate to fit between the lateral face of the otic capsule and the dorsal edge of the quadrate; the wider head rests on a posterodorsal concavity on the otic capsule, and the smaller fits a lower, more anterior pit. Laterally the stapes carries a short, broad process that probably made contact with a dorsally placed tympanic membrane. Thus the bone was a hyomandibular in the sense that it articulated with the quadrate, but it may also have served as a stapes in sound-transmission. It contains no visible canal or foramen.

MANDIBLE

The crushed inner surface of the posterior part of the left mandible and most of the external surface of the right mandible are preserved in close proximity. Although the whole length of the tooth-bearing margins is missing, some parts of six elements of the right mandible can be seen. The pattern of sutures and the general contour closely resemble those of *Megalichthys* (Watson, 1926, Figs. 37, 38) and other known Rhipidistia.

The anteroposterior length of the mandible is about 23.8 mm., and the depth is 3.8 mm. The dentary extends approximately 17.6 mm. back from the symphysis, and its greatest width is probably 2.0 mm. Its lower edge meets all the other lateral bones of the jaw. The splenial and postsplenial form the curved anteroventral half of the jaw for a distance of about 9.0 mm. The fragmented articular, on the posterior end of the jaw, is 4.0 mm. long and 2.0 mm. deep, exhibiting a broken upper edge; presumably the surface for articulation with the quadrate was a shallow concavity, above the end of the articular.

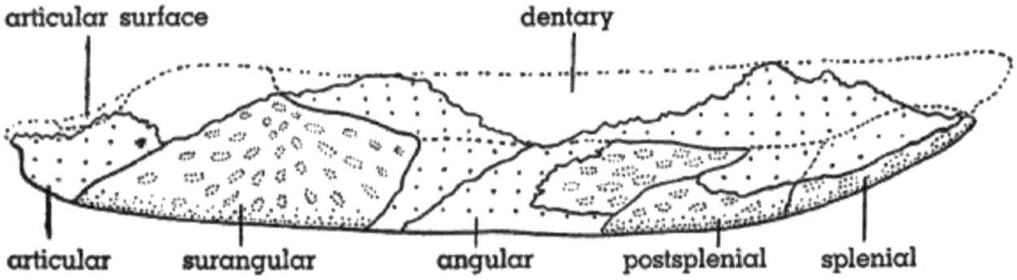

Fig. 6. *Hesperoherpeton garnettense* Peabody. Right mandible, lateral view, KU 10295, x 4. External surfaces are pitted; broken surfaces are coarsely stippled.

VERTEBRAE

(Fig. 7)

The vertebrae that are visible from a lateral view are crushed and difficult to interpret. It is possible, nevertheless, to see that the trunk vertebrae resemble those of Ichthyostegalia (Jarvik, 1952, Fig. 13 A, B), except that the pleurocentra are much larger. A few parts of additional vertebrae can be seen, but they are so scattered that it is impossible to be sure of their original location. Therefore comparisons between different regions cannot yet be made.

The U-shaped intercentrum encloses the notochord and occupies an anteroventral position in the vertebra. Anteriorly, each intercentrum articulates with the pleurocentra of the next preceding vertebra by slightly concave surfaces. Dorsolaterally there is an articular surface for the capitulum of the rib.

The two pleurocentra of each vertebra are separate ventrally as well as dorsally, but form thin, broad plates of about the same height as the notochord. The lateral surface appears to be depressed, allowing, perhaps, for movement of the rib. Above each pleurocentrum, on the lateral surface of the neural arch, there is a short diapophysis for articulation with the tuberculum of the rib.

The margin of the neural spine is convex anteriorly and concave posteriorly, the tip reaching a point vertically above the postzygapophysis. The prezygapophysis of each vertebra articulates with the preceding postzygapophysis by a smooth dorsal surface. One nearly complete neural arch shows (Fig. 7 B) a pit above the neural canal, clearly corresponding to the canal for a dorsal ligament shown by Jarvik in *Ichthyostega*. Indeed this view of the neural arch and intercentrum together brings out the striking resemblance between the vertebrae of *Hesperoherpeton* and those of the Ichthyostegids. The rounded intercentrum in both is an incomplete ring enclosing the notochordal canal.

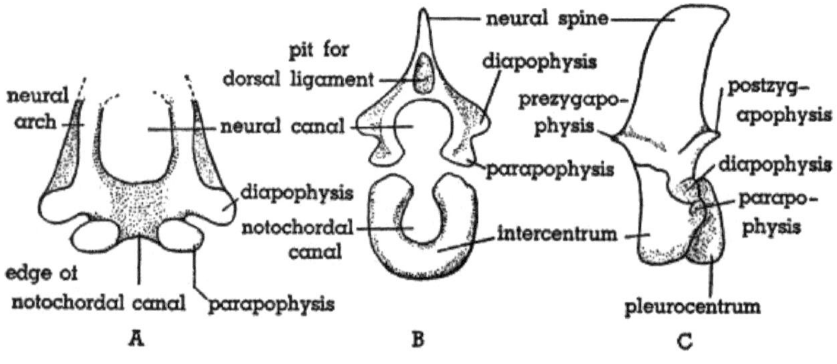

Fig. 7. *Hesperoherpeton garnettense* Peabody. A, End view of incomplete vertebra, probably near anterior end of column. B, Neural arch and intercentrum in end view, showing probable association. C, Left lateral view of trunk

vertebra. All figures: KU 10295, x 4.

TABLE 1.--AVERAGE MEASUREMENTS OF THE TRUNK VERTEBRAE (in mm.).

NUMBERS IN PARENTHESES INDICATE THE NUMBER OF PIECES AVAILABLE FOR MEASURING

Parts	Ant.-post.	Dors.-vent.	Transv. width
Neural spine	1.5 (3)	3.0 (3)	--
Neural spine and arch	2.0 (4)	4.5? (4)	--
Neural canal	2.0 (4)	2.0 (1)	

1.0 (1)

Intercentrum

1.5 (5)

3.5 (4)

3.0 (1)

Pleurocentrum

1.5 (3)

3.0 (2)

--

The shape, in end view, of a partly preserved neural arch (Fig. 7 A) seems to account for the incompleteness of the intercentrum just mentioned; the ventral edge of the arch is emarginate in such a way as to fit the dorsal surface of the notochord. The dorsal portion of this neural arch is not present (either broken or not yet ossified), but the opening of the neural canal is comparable in width to the foramen magnum. Hence this vertebra may be one of the most anterior in the column. In comparison with the trunk vertebrae seen farther posteriorly it appears that there may be a progressive ossification of neural arches toward their dorsal ends, and of intercentra around the notochord, with probable fusion of the intercentra and neural arches in the posterior part of the trunk. The notochord seems to have been slightly constricted by the intercentra, but not interrupted.

RIBS

The proximal ends of the ribs expand dorsoventrally to a width approximately four times that of their slender shafts. The tuberculum and capitulum on each of the trunk ribs are separated only by a shallow concavity. These two articular surfaces are so situated that the rib must tilt downward from the horizontal plane. The shaft flares terminally in some ribs, and the distal end is convex. Ribs in the trunk region differ little if any in size. Five that can be measured vary in length from 5.0 to 7.0 mm. One short, bent rib 3.5 mm. long perhaps is sacral or caudal.

PECTORAL GIRDLE

(Figs. 8, 9, 10)

The right scapulocoracoid is almost complete, and the left one is present but partly broken into three pieces, somewhat pushed out of position. With the advantage of this new material, we may comment on the scapulocoracoid of *H. garnettense* as described by Peabody (1958). In size and contour, the slight differences between the type (KU 9976) and the new skeleton (KU 10295) are considered to be no more than individual variation. We have redrawn the type (Fig. 8) in order to show the resemblances more clearly.

The small sections that were missing from the type are present in KU 10295. The jagged edge directly posterior to the area occupied by the neural arch in the type extends 0.5 mm. farther back in our specimen. The angle formed between the recurved dorsal ramus and the edge of the ventral flange is seen in our specimen to be less than 90deg. The glenoid fossa, appearing as a concave articular surface for the cap of the humerus, was in part covered by cartilage and shows as "unfinished" bone (Peabody, 1958, p. 572); this area is more oval than triangular, as Peabody thought. The obstruction of a clear view of this part of the type is the result of the accidental position of a neural arch. The raised portion immediately dorsal to the glenoid fossa exhibits an unfinished surface, suggesting the presence of either cartilage or a ligament.

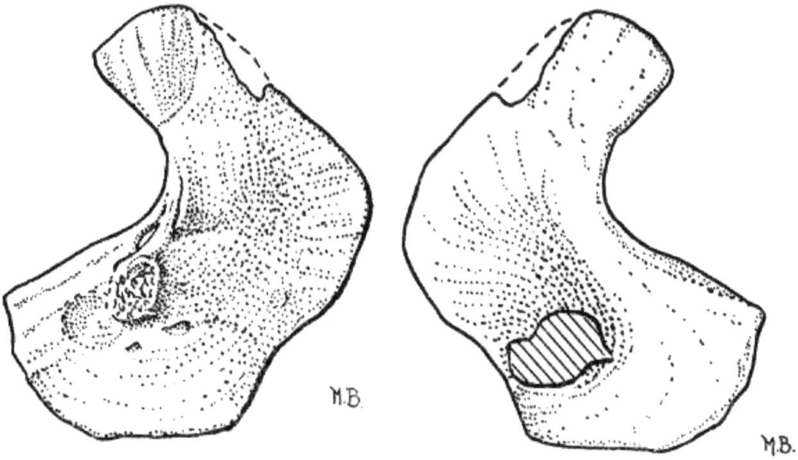

Fig. 8. *Hesperoherpeton garnettense* Peabody. Type specimen redrawn. Right scapulocoracoid in external view (at left), and internal view (at right). KU 9976, x 4.

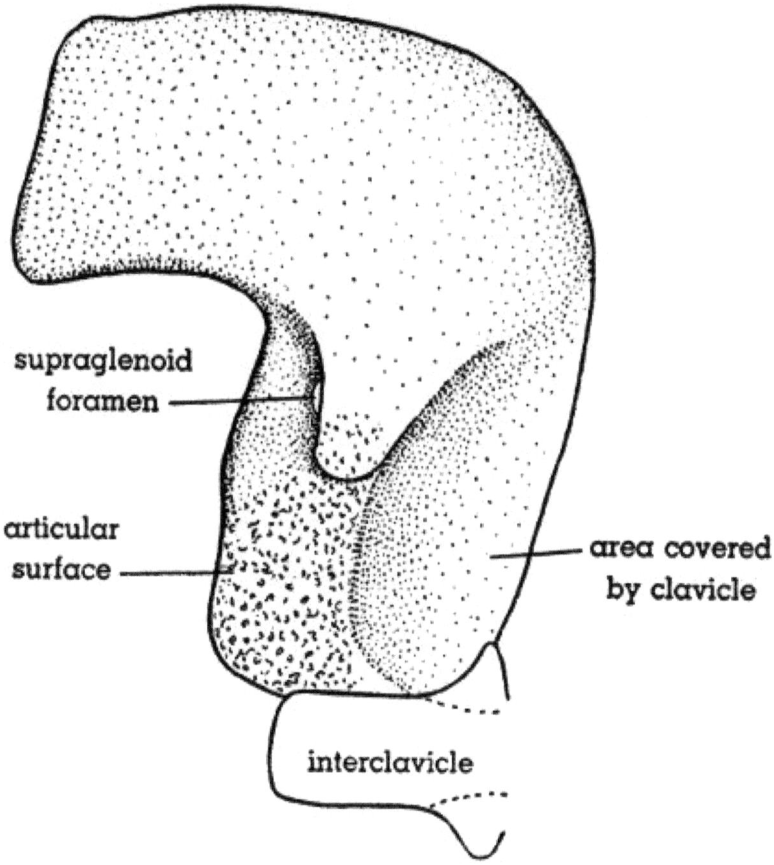

9

Fig. 9. *Hesperoherpeton garnettense* Peabody. Right scapulocoracoid in external view, showing part of interclavicle, and position occupied by clavicle. The specimen is flattened and lies entirely in one plane. KU 10295, x 4.

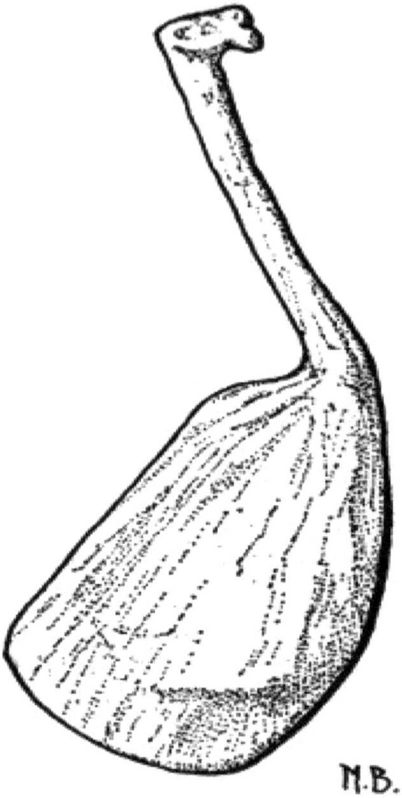

10

Fig. 10. ***Hesperoherpeton garnettense*** **Peabody. Right clavicle in external view. Anterior edge to right. KU 10295, x 4.**

The right clavicle is complete, and resembles a spoon having a slender handle. The dorsal tip of the handle is L-shaped. The expanded ventral part is convex externally, and rested upon the anteroventral surface of the scapulocoracoid. The lateral edge next to the "stem" is distinctly concave, abruptly becoming similar in contour to the opposite edge, and giving the impression of an unsymmetrical spoon. The left clavicle is present in scattered fragments, its dorsal hooklike end being intact.

The posterior end of the interclavicle lies in contact with the right scapulocoracoid. There are short lateral processes at the point where the interclavicle was overlapped by the clavicles, but we cannot be sure of the extent of this bone anteriorly or posteriorly.

The presumed left cleithrum, a long rectangle, is approximately equal in length to the rodlike stem of the clavicle, and is about as wide as the dorsal L-shaped tip of the clavicle. The posterior end of the cleithrum presumably met the tip of the clavicle, while the rest of it was directed anteriorly and a little dorsally. There seems to be a small articular surface near the anterior extremity which suggests the presence of a supracleithrum. The upper border of the cleithrum is slightly convex and the lower concave.

FORELIMB

(Fig. 11)

The left forelimb is the only one present and appears to be nearly complete, although the elements are scattered almost at random. The only parts of the forelimb known to be missing are two subterminal and two terminal phalanges, probably of the first and third digits, and the proximal end of the second metacarpal. The smooth and relatively flat surfaces suggest an aquatic rather than terrestrial limb; only the proximal half of the humerus bears any conspicuous ridges or depressions. As we restore the skeleton of the limb, several features are remarkable: The humerus, ulna, and ulnare align themselves as the major axis of the limb, each carrying on its posterior edge a process or flange comparable to those in the axial series of a rhipidistian fin. The remaining elements take positions comparable to the diagonally placed preaxial radials in such a fin. The digits appear to have been short, perhaps with no more than two phalanges. There is only one row of carpals present (the proximal row of other tetrapods). A second and third row would be expected in primitive Amphibia; if they existed in *Hesperoherpeton* they must either have been wholly cartilaginous or washed away from the specimen. Neither of these alternatives seems at all likely to us in view of the well-ossified condition of the elements that are present, and the occurrence of both the proximal carpals and the metacarpals. The space available for metacarpals probably could not have contained more than the four that are recognized.

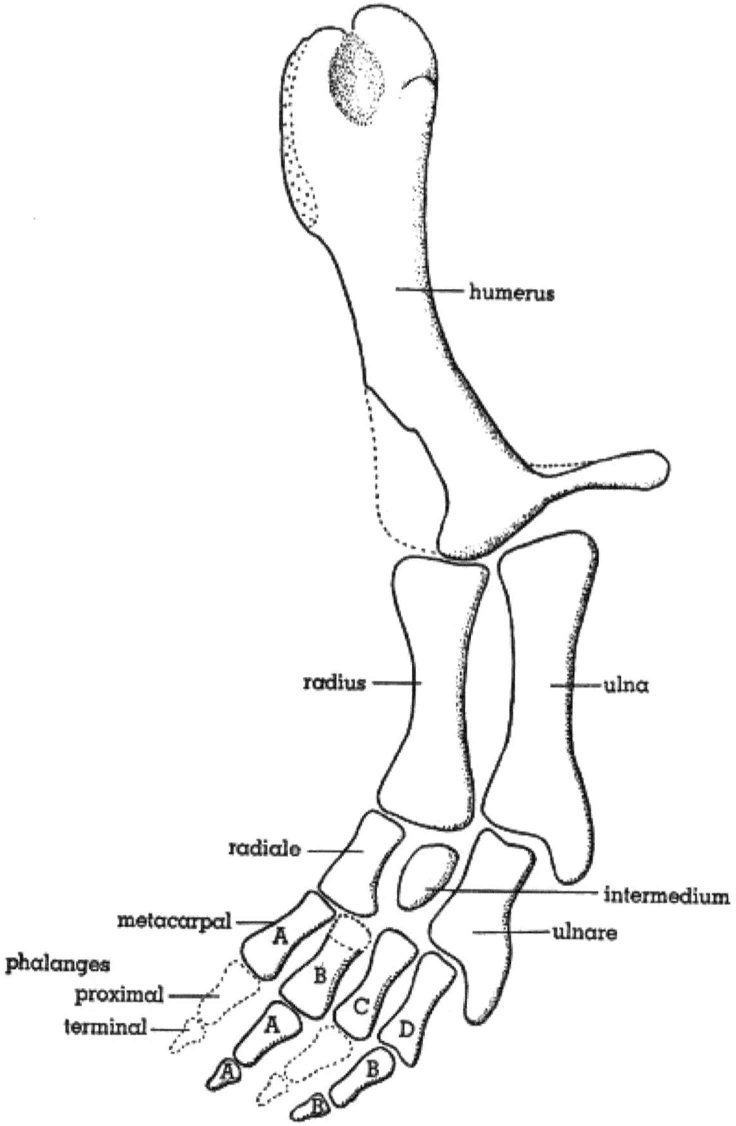

Fig. 11. ***Hesperoherpeton garnettense*** **Peabody. Left forelimb, showing characters of both a crossopterygian fin and an amphibian foot. KU 10295, x 4.**

The proximal end of the humerus is more rounded anteriorly than posteriorly, and has a thin articular border that bore a cartilaginous cap as the primary surface for articulation with the scapulocoracoid. Although the unfinished surface of the head extends down the anterior margin about a third the length of the humerus, the shaft has been broken and so twisted that the distal part is not in the same plane as the proximal. Immediately posterior to the cartilaginous cap is a round, deep notch bordered posteriorly by the dorsal process of the head.

The shaft is longer and narrower than would be anticipated in a primitive amphibian limb (cf. Romer, 1947). The distal end bears two surfaces for articulation with the radius and ulna. The full extent of the former surface was not determined because the more anterior part of the expanded end is represented only by an impression. The surface nearest the ulna was partially rounded for articulation with that element, the remaining posterior edge being broadly concave. The most striking feature of the humerus is a slender hooklike process on the posterior edge near the distal end, probably homologous with (1) the posterior flange on the "humerus" in Rhipidistia, and (2) the entepicondyle of the humerus in *Archeria* (Romer, 1957) and other tetrapods.

The radius is about the same width proximally as distally. The curvature of the shaft is approximately alike on both sides. Distally the surface is rounded for articulation with the radiale and perhaps the intermedium.

The proximal end of the ulna is similar to that of the radius but is slightly larger. Posteriorly, there is a short, broad expansion resembling the entepicondyle of the humerus, and even more nearly like the postaxial flanges in a crossopterygian fin.

The ends of the radiale are expanded and rounded, the entire bone being approximately twice as long as wide. The three sides of the intermedium are similarly convex. The surface of this bone is unfinished, showing that it must have been embedded in cartilage. The ulnare is conspicuously similar to the ulna in bearing a posterior hooklike expansion, and is larger than the radiale.

The four metacarpals are slightly expanded proximally and distally. Although measurements of length and width are tabulated below (Table 2), we are not certain of the sequence of these bones in the row.

The dimensions of the two proximal phalanges are alike. The shape of these elements is similar to that of the metacarpals. The two terminal phalanges are somewhat triangular in shape, the lateral edges being concave and the proximal convex.

TABLE 2.--APPROXIMATE MEASUREMENTS OF THE FORELIMB (in mm.)

Element

Dimensions

Length

Width

Proximal

Midway

Distal

Humerus

16.0

5.0

2.0

7.5?

Radius

9.0

4.0

1.5

3.5

Ulna

8.5

4.5

1.5

3.5

Radiale

3.0

2.0

1.5

2.0

Intermedium

1.5

--

2.0

--

Ulnare

3.5

2.0

2.0

2.5

Metacarpal A

4.5

2.5

1.0

2.0

Metacarpal B

4.5

3.0?

1.5

2.5

Metacarpal C

4.0

2.0

1.5

2.0

Metacarpal D

3.5

2.5

1.0

1.5

Proximal Phalanx A

2.0

1.5

1.0

1.5

Proximal Phalanx B

2.0

1.5

1.0

1.5

Terminal Phalanx A

1.5

1.5

1.0

1.0

Terminal Phalanx B

1.5

1.5

1.0

1.0

COMPARISONS AND DISCUSSION

Apparently primitive rhipidistian characters in *Hesperoherpeton* are: Braincase in two sections, posterior one containing an expanded notochordal canal; lateral series of mandibular bones closely resembling that of *Megalichthys*, as figured by Watson (1926); tabular having long process probably articulating with pectoral girdle; lack of movement between head and trunk correlated with absence of occipital condyle; sensory pits present on frontal and squamosal.

Although we are unable to separate, by sutures, the vomers from the palatines, the palatal surface of these bones and of the pterygoids is studded by numerous small teeth, as in Rhipidistia (Jarvik, 1954) and some of the early Amphibia (Romer, 1947). The stapes apparently reaches the quadrate, and could therefore serve in hyostylic suspension of the upper jaw.

The pectoral limb has an axial series of bones carrying hooklike flanges on their posterior edges. The other bones of the limb show little modification of form beyond the nearly flat, aquatic type seen in Rhipidistia. No distinct elbow or wrist joints are developed.

Characters of *Hesperoherpeton* common to most primitive Amphibia, in contrast with Crossopterygii, are: Nares separated from edge of jaw; stapes having external process that may have met a tympanic membrane, thus giving the bone a sound-transmitting function. Apparently none of the opercular series was present.

There are two large palatal teeth, slightly labyrinthine in character, adjacent to each internal naris. The scapulocoracoid, as shown by Peabody (1958), is Anthracosaurian in structure, as are the long-stemmed clavicles. The limbs have digits rather than fin-lobes, although the digital number apparently is four and the number of bones in the manus is less than would be expected in a primitive amphibian. The vertebrae are similar to those of Ichthyostegids, as described by Jarvik (1952), except that the pleurocentra are much larger.

In addition to this remarkable combination of crossopterygian and amphibian characters, *Hesperoherpeton* is specialized in certain features of the skull. The orbits are much enlarged, probably in correlation with the diminutive size of the animal, and this has been accompanied by loss of several bones. The frontal and squamosal nearly meet each other, and both form part of the rim of the orbit. The bones of the posterior part of the dermal roof are greatly reduced, and there is none behind the squamosal except the projecting tabular; there is no indication of quadratojugal, jugal, intertemporal or postparietal. The foramen magnum is enormous. The external surfaces of the bones of the skull are nearly smooth.

Is it possible that the "primitive" and "specialized" features of this animal are actually larval? Are they not just the kind of characters that would be expected in an immature, aquatic embolomere of Pennsylvanian time? For several reasons we do not think this is the case. Except for the anterior part of the braincase, there is no indication that the skeleton was not well ossified. The postaxial processes on the humerus, ulna and ulnare could scarcely have been larval features only, since they are so clearly homologous with those in adult Rhipidistia; a larval limb should indeed be simple, but its simplicity is unlikely to involve paleotelic adult characters. The scapulocoracoid of our specimen is of practically the same shape and size as that in the only other known individual, the type; this would be probable if both were adults, but somewhat less likely if they were larvae of a much larger animal. The form of the stapes, tabular and otic notch suggest a functional tympanic membrane, which could not have occurred in a gill-breathing larva. On the other hand, an adult animal of pigmy size might be expected to have large orbits, large otic capsules and a large foramen magnum.

We conclude that *Hesperoherpeton* lived and sought food in the weedy shallows at the margin of a pond or lagoon, and that for much of the time its head was partly out of water (Fig. 12). The animal could either steady itself or crawl around by means of the paddlelike limbs, but these probably could not be used in effective locomotion on land. Like the Ichthyostegids, it probably swam by means of a fishlike tail.

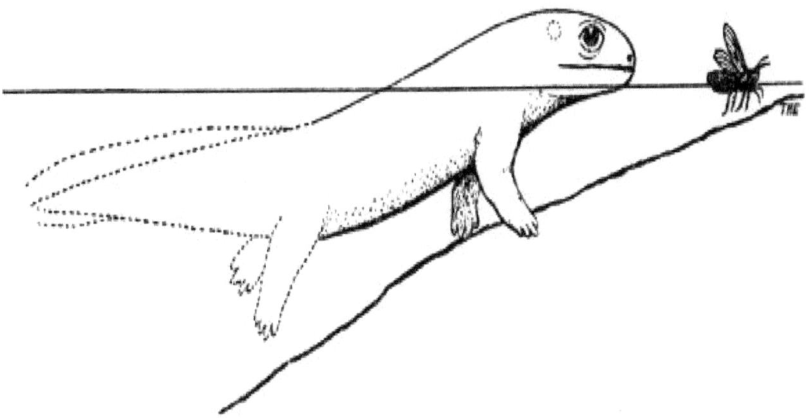

Fig. 12. *Hesperoherpeton garnettense* Peabody. Probable appearance in life. x 0.5.

TAXONOMY

Evidently *Hesperoherpeton* is a small, lagoon-dwelling survivor of the Devonian forms that initiated the change from Crossopterygii to Amphibia (Jarvik, 1955). It shows, however, that this transition did not affect all structures at the same time, for some, as the braincase with its notochordal canal, the mandibular bones and axial limb bones, are unchanged from the condition normal for the Rhipidistia, but most other characters are of amphibian grade. To express these facts taxonomically requires that *Hesperoherpeton* be removed from the family Cricotidae, suborder Embolomeri, order Anthracosauria, and placed in a new order and family of labyrinthodont Amphibia.

ORDER PLESIOPODA

(*plesios*, Gr., near, almost; *podos*, Gr., foot)

Labyrinthodontia having limbs provided with digits, but retaining posterior flanges on axial bones as in Rhipidistia, without joint-structure at elbow and wrist essential for terrestrial locomotion; neurocranium having separate otico-occipital section, large notochordal canal, no occipital condyle, as in Rhipidistia; nares separate from rim of mouth; pectoral girdle anthracosaurian; vertebrae having U-shaped intercentrum and paired, but large, pleurocentra.

Probably associated with the characters of the order, as given above, are the connection of pectoral girdle with skull, and the presence of a tympanic membrane, the stapes functioning in both sound-transmission and palatoquadrate suspension.

FAMILY HESPEROHERPETONIDAE

Orbits and foramen magnum unusually large in correlation with reduced size of animal; squamosal forming posterior margin of orbit; circumorbital series absent (except for postorbital); sensory pits on squamosal and frontal.

Characters defining the family are evidently the more specialized cranial features, which probably evolved during Mississippian and early Pennsylvanian times.

The definition of the genus and species may be left to rest upon Peabody's (1958) original description and the present account, until the discovery of other members of the family gives reason for making further distinctions.

THE END